SURVEYING MATHEMATICS MADE SIMPLE

An original Book by

Jim Crume P.L.S., M.S., CFedS

Co-Authors
Cindy Crume
Bridget Crume
Troy Ray R.L.S.
Mark Sandwick L.S.I.T.

PRINTED EDITION

*****

PUBLISHED BY:

Jim Crumo P.L.S., M.S., CFedS

Circular Curves

Book 4 of this Math-Series

Copyright 2013 © by Jim Crume P.L.S., M.S., CFedS

All Rights Reserved

First publication: November, 2013

Printed by CreateSpace

Available on Kindle and other devices

TERMS AND CONDITIONS

The content of the pages of this book is for your general information and use only. It is subject to change without notice.

Neither we nor any third parties provide any warranty or guarantee as to the accuracy, timeliness, performance, completeness or suitability of the information and materials found or offered in this book for any particular purpose. You acknowledge that such information and materials may contain inaccuracies or errors and we expressly exclude liability for any such inaccuracies or errors to the fullest extent permitted by law.

Your use of any information or materials in this book is entirely at your own risk, for which we shall not be liable. It shall be your own responsibility to ensure that any products, services or information available in this book meet your specific requirements.

This book may not be further reproduced or circulated in any form, including email. Any reproduction or editing by any means mechanical or electronic without the explicit written permission of Jim Crume is expressly prohibited.

Table of Contents

INTRODUCTION..4

CIRCULAR CURVE..6

REVERSE CURVE..16

COMPOUND CURVE..18

PRACTICAL EXAMPLE..22

SOLUTIONS TO EXAMPLES..................................25

ABOUT THE AUTHOR..33

INTRODUCTION

Straight forward Step-by-Step instructions.

This book is just one part in a series of digital and printed editions on Surveying Mathematics Made Simple. The subject matter in this book will utilize the methods and formulas that are covered in the books that precede it. If you have not read the preceding books, you are encouraged to review a copy before proceeding forward with this book.

For a list of books in this series, please visit:

http://www.cc4w.net/ebooks.html

Prerequisites for this book:

A basic knowledge of geometry, algebra and trigonometry is required for the explanations shown in this book.

Book 1 - **Bearings and Azimuths** - How to add bearings and angles, subtract between bearings, convert from degrees-minutes-seconds to decimal degrees, convert from decimal degrees to degrees-minutes-seconds, convert from bearings to azimuths and convert from azimuths to bearings.

Definitions:

Circular Curve (a.k.a. Simple Curve): Is a curve with a constant radius. All points on the curve are equal distance from the center of the circle.

Tangent: A straight line that touches a given curve at one point only that is perpendicular to a straight line to the center of the circle.

A circular curve is often referred to as a horizontal curve which is misleading since a spiral curve is also a horizontal curve. Book 6 - "Spiral Curves" details the common highway and railroad spiral curves.

CIRCULAR CURVE

Figure 1 shows the various components for a circular curve. It is important that you become familiar with these components. They will be referenced throughout this book.

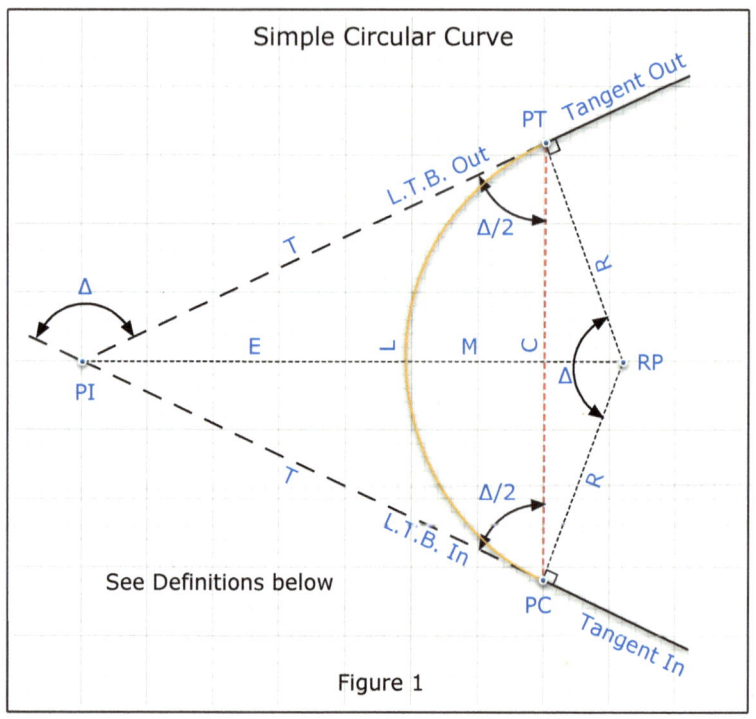

Figure 1

Figure 2 shows the relationship of the Degree of Curvature (D) with a 100' arc length (L) which is known as the "Arc Definition". The Arc Definition is used on highway curves and curves along boundary lines.

Figure 3 shows the relationship of the Degree of Curvature (D) with a 100' chord distance (C) which is known as the "Chord Definition". The Chord definition is used mainly on railroad curves.

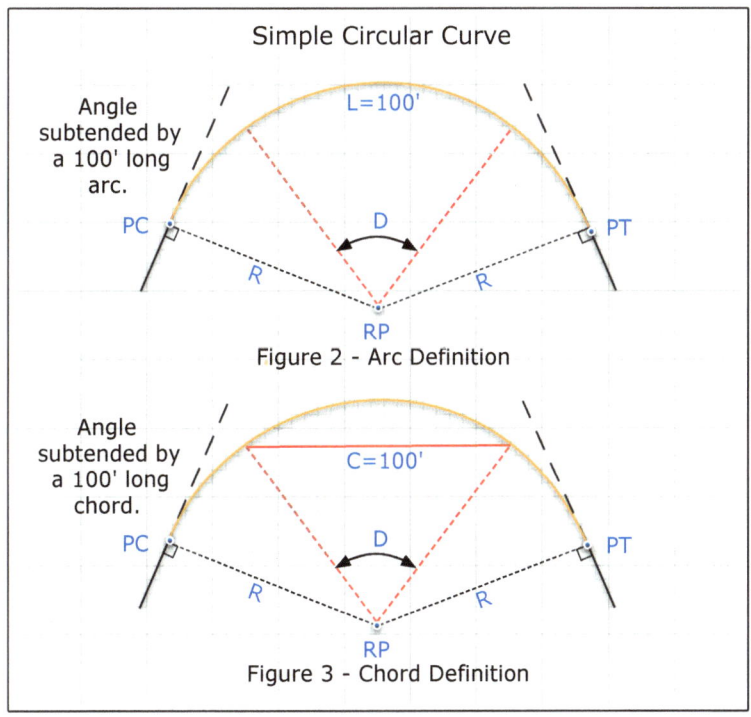

Definitions:

D = Degree of Curvature (Arc and Chord Definition)

R = Radius

RP = Radius Point

PC = Point of Curvature

PI = Point of Intersection

PT = Point of Tangency

Δ = Delta or Central Angle (RT or LT)

T = Tangent distance

C = Chord distance

L = Arc or Curve Length

E = External distance

M = Middle Ordinate

RT = Curve to the Right (Clockwise)

LT = Curve to the Left (Counter Clockwise)

L.T.B. = Local Tangent Bearing is a line perpendicular from a line drawn from the radius point to the connecting PC or PT. (Used for non-tangent curves, Point of Reverse Curve and Point of Compound Curve)

A non-tangent curve = Where the course or line coming into the PC or leaving the PT is not tangent (perpendicular) to the radial line to the Radius Point.

Radial Line = A straight line from the radius point of a circular curve to a designated point on the curve.

Common occurrences for Circular Curves:

The Δ angle is always equal to the deflection angle at the PI.

The deflection angle at the PC and PT as measured from the tangent line to the chord line is always one half of the Δ angle. (Δ/2)

The tangent length (T) for the "Tangent In" is always equal to the "Tangent Out" distance.

Formulas:

Arc Definition - Figure 2 (Highway Circular Curve)

$D = 5729.57795 / R$

$R = 5729.57795 / D$

Chord Definition - Figure 3 (Railroad Circular Curve)

$D = ArcSin(50 / R) * 2$

$R = 50 / Sin(D / 2)$

$L = 100 * Δ / D$

$Δ = L * D / 100$

Common to both curve definitions:

$R = (180 * L) / (Δ * π)$

$R = T / Tan(Δ / 2)$

$R = C / (2 * Sin(Δ / 2))$

$L = (Δ * R * π) / 180$

$L = (π * C * Δ) / (360 * Sin(Δ / 2))$

$$L = (\pi * T * \Delta) / (180 * \tan(\Delta / 2))$$

$$C = 2 * R * \sin(\Delta / 2)$$
$$C = (360 * L * \sin(\Delta / 2)) / (\pi * \Delta)$$
$$C = 2 * T * \cos(\Delta / 2)$$
$$C = 2 * R * \sin((90 * L) / (\pi * R))$$
$$C = (2 * T * R) / \sqrt{(R^2 + T^2)}$$

$$\Delta = (180 * L) / (R * \pi)$$
$$\Delta = 2 * \arctan(T / R)$$
$$\Delta = 2 * \arcsin(C / (2 * R))$$

$$T = R * \tan(\Delta / 2)$$
$$T = (180 * L * \tan(\Delta / 2)) / (\pi * \Delta)$$
$$T = C / (2 * \cos(\Delta / 2))$$
$$T = R * \tan((90 * L) / (\pi * R))$$
$$T = (R * C) / (\sqrt{((4 * R^2) - C^2)})$$

$$M = R * (1 - \cos(\Delta / 2))$$
$$M = (180 * L * (1 - \cos(\Delta / 2))) / (\pi * \Delta)$$
$$M = (C * \tan(\Delta / 4)) / 2$$
$$M = T * (1 - \cos(\Delta / 2)) / \tan(\Delta / 2)$$

$$E = (R / \cos(\Delta / 2)) - R$$
$$E = (180 * L * (1 - \cos(\Delta / 2))) / (\pi * \Delta * \cos(\Delta / 2))$$
$$E = (C * \tan(\Delta / 4)) / (2 * \cos(\Delta / 2))$$
$$E = T * \tan(\Delta / 4)$$

Chord Bearing = Tangent In or L.T.B. In (+/-) (Δ / 2)

Tangent Out = Tangent In or L.T.B. In (+/-) Δ

Notes

Example 1:

Refer to Figure 1 for curve nomenclature.

Given:

Δ = 10°31'20" RT

R = 2864.78898

Tangent In = N50°20'45"W

Solve for the following elements:

T = R * Tan(Δ / 2)

T = 2864.78898 * (Tan(10°31'20" / 2)

T = **263.79738**

L = (Δ * R * π) / 180

L = (10°31'20" * 2864.78898 * π) / 180

L = **526.11111**

C = 2 * R * Sin(Δ / 2)

C = 2 * 2864.78898 * Sin(10°31'20" / 2)

C = **525.37210**

Chord Bearing = Tangent In - (Δ / 2)

Chord Bearing = N50°20'45"W - (10°31'20" / 2)

Chord Bearing = **N45°05'05"W**

Tangent Out = Tangent In - Δ

Tangent Out = N50°20'45"W - 10°31'20"

Tangent Out = **N39°49'25"W**

Example 2:

Given:

T = 1187.69109
Δ = 63°45'10" LT
Tangent In = N35°23'11"E

Solve for the following elements:
R = T / Tan(Δ / 2)
R = 1187.69109 / Tan(63°45'10" / 2)
R = **1909.85940**
L = (π * T * Δ) / (180 * Tan(Δ / 2))
L = (π * 1187.69109 * 63°45'10") / (180 * Tan(63°45'10" / 2))
L = **2125.09268**
C = 2 * T * Cos(Δ / 2)
C = 2 * 1187.69109 * Cos(63°45'10" / 2)
C = **2017.14932**
Chord Bearing = Tangent In - (Δ / 2)
Chord Bearing = N35°23'11"E - (63°45'10" / 2)
Chord Bearing = **N3°30'36"E**
Tangent Out = Tangent In - Δ
Tangent Out = N35°23'11"E - 63°45'10"
Tangent Out = **N28°21'59"W**

Note: Rounding error is dependent upon the number of decimal places that are utilized. It is recommended that at least 5 decimal places be used for all calculations then round the final answer as needed.

All angles must be converted to Decimal Degrees prior to performing trigonometric operations. See Book 1 - "Bear-

ings and Azimuths" for methods on converting Degrees-Minutes-Seconds to Decimal Degrees and vice versa. Also see Book 1 for adding and subtracting bearings and angles.

Examples 1 & 2 walked you through the steps to calculate the unknown values for two circular curve solutions. You can solve almost all unknown elements for a circular curve with just two known values which is usually a Radius (R) and the Delta or Central Angle (Δ).

Now that you have a couple of examples to follow, try solving the missing elements for the following curve given the Delta (Δ) and Arc Length (L).

Example 3:

Given:

$\Delta = 30°21'13"$ RT

L = 1011.79055

Tangent In = S29°56'23"W

Solve for the following elements:

C = ????.?????

R = ????.?????

T = ???.?????

D (Arc definition) = ?°??'??"

D (Chord definition) = ?°??'??"

Chord Bearing = x??°??'??"x

Tangent Out = x??°??'??"x

The solution can be found at the end of the book.

NOTES

REVERSE CURVE

Definitions:

A Reverse Curve is two circular curves having a common tangent, the curves lying on opposite sides of the common tangent.

PRC = Point of Reverse Curve

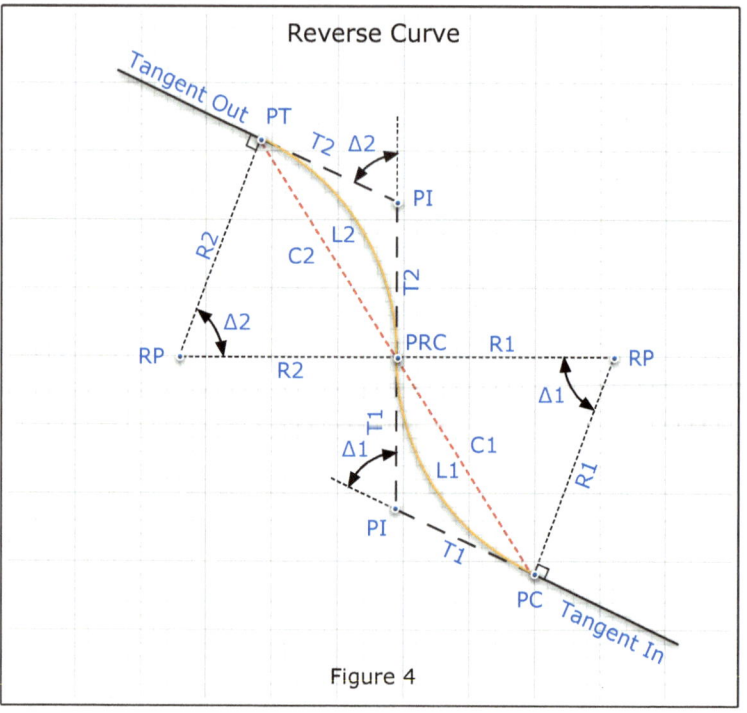

Figure 4

Additional Formula:

Tangent Out = Tangent In (+/-) (Δ1 - Δ2)

Example 4 - Reverse Curve:

Refer to Figure 4 for the following example:

Given:

$R_1 = 1200.00000$

$R_2 = 900.00000$

$\Delta_1 = 63°45'10''$ RT

$\Delta_2 = 66°33'12''$ LT

Tangent In = N55°23'11"W

Solve for the following elements:

T_1 = ????.?????

T_2 = ????.?????

L_1 = ????.?????

L_2 = ????.?????

L.T.B. at PRC = x??°??'??"x

Tangent Out = x??°??'??"x

C_1 = ????.?????

Chord Bearing 1 = x??°??'??"x

C_2 = ????.?????

Chord Bearing 2 = x??°??'??"x

The solution can be found at the end of the book

COMPOUND CURVE

Definitions:

A Compound Curve is two circular curves of different radius which are tangent at one point with both curves lying on the same side of the common tangent.

PCC = Point of Compound Curve

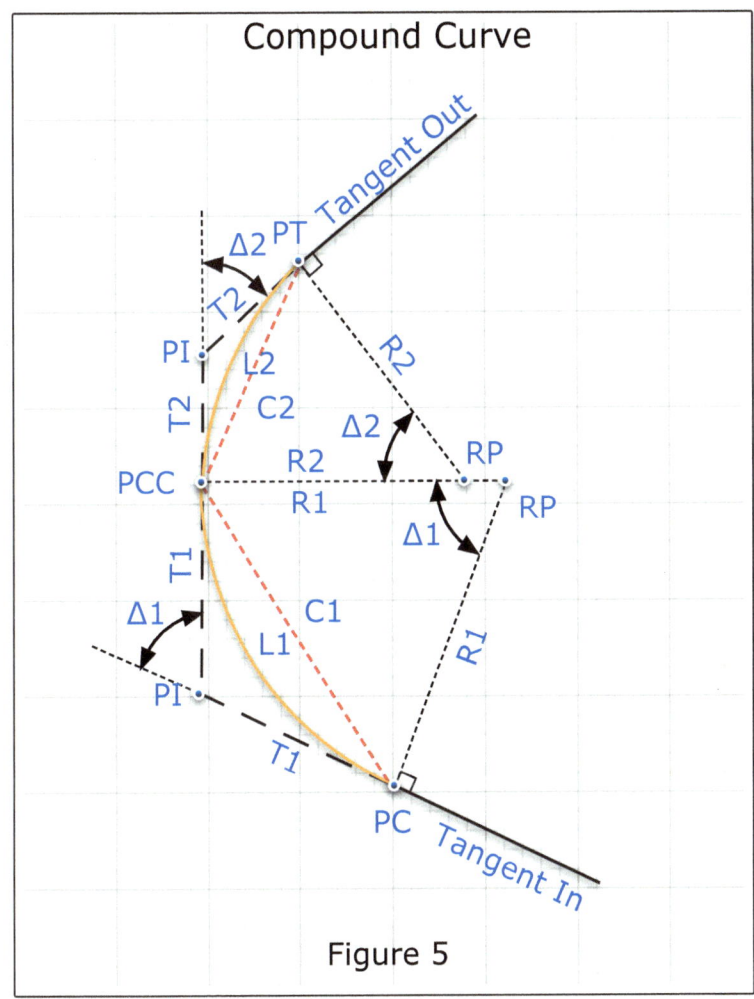

Figure 5

Additional Formula:

Tangent Out = Tangent In (+/-) (Δ1 + Δ2)

Example 5 - Compound Curve:

Refer to Figure 5 for the following example:

Given:

$R_1 = 1000.00000$

$R_2 = 700.00000$

$\Delta_1 = 63°45'10''$ RT

$\Delta_2 = 40°34'32''$ RT

Tangent In = N53°25'41"W

Solve for the following elements:

T_1 = ????.?????

T_2 = ????.?????

L_1 = ????.?????

L_2 = ????.?????

L.T.B. at PCC = x??°??'??"x

Tangent Out = x??°??'??"x

C_1 = ????.?????

Chord Bearing 1 = x??°??'??"x

C_2 = ????.?????

Chord Bearing 2 = x??°??'??"x

The solution can be found at the end of the book.

NOTES

PRACTICAL EXAMPLE

It is common for boundary lines to follow curved segments along controlling lines. Figure 6 illustrates a boundary line connected to a curved controlling line then following the curve for a distance then exiting the curved segment. In these instances, L.T.B.'S, Chord Bearings or Radial lines are required for legal descriptions in order to describe the curved segment properly.

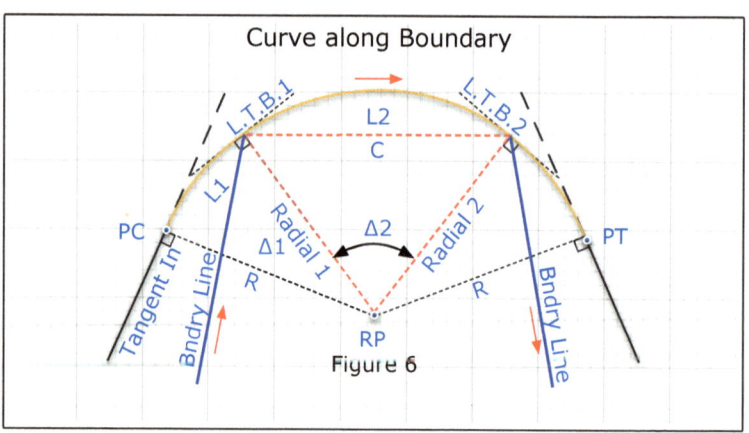

Figure 6

Example 6:

Refer to Figure 6 for the following example.

Given:

R = 2864.78898

L1 = 85.34'

L2 = 150.11'

Tangent In = N10°20'45"E

Solve for the following elements:

L.T.B. 1 = x??°??'??"x
L.T.B. 2 = x??°??'??"x
Δ1 = ??°??'??" RT
Δ2 = ??°??'??" RT
C = ????.?????
Chord Bearing = x??°??'??"x
Radial Bearing 1 = x??°??'??"x
Radial Bearing 2 = x??°??'??"x

The solution can be found at the end of the book.

NOTES

SOLUTIONS TO EXAMPLES

Solution for Example 3 - Circular Curve:

Given:

Δ = 30°21'13" RT

L = 1011.79055

Tangent In = S29°56'23"W

Solve for the following elements:

C = (360 * L * Sin(Δ / 2)) / (π * Δ)

C = (360 * 1011.79055 * Sin(30°21'13" / 2)) / (π * 30°21'13")

C = **1000.00008**

R = (180 * L) / (Δ * π)

R = (180 * 1011.79055) / (30°21'13" * π)

R = **1909.86595**

T = (180 * L * Tan(Δ / 2)) /(π * Δ)

T = (180 * 1011.79055 * Tan(30°21'13" / 2) / (π * 30°21'13")

T = **518.06896**

D = 5729.57795 / R (Arc definition)

D = 5729.57795 / 1909.86595

D = **3°00'00"**

D = ArcSin(50 / R) * 2 (Chord definition)

D = ArcSin(50 / 1909.86595) * 2

D = **3°00'01"**

Chord Bearing = Tangent In + (Δ / 2)
Chord Bearing = S29°56'23"W + (30°21'13" / 2)
Chord Bearing = **S45°07'00"W**
Tangent Out = Tangent In + Δ
Tangent Out = S29°56'23"W + 30°21'13"
Tangent Out = **S60°17'36"W**

Solution for Example 4 - Reverse Curve:

Refer to Figure 4 for the following example:
Given:
R1 = 1200.00000
R2 = 900.00000
Δ1 = 63°45'10" RT
Δ2 = 66°33'12" LT
Tangent In = N55°23'11"W

Solve for the following elements:
T = R * Tan(Δ / 2)
T1 = 1200.00000 * Tan(63°45'10" / 2)
T1 = **746.24829**
T2 = 900.00000 * Tan(66°33'12" / 2)
T2 = **590.66497**
L = (Δ * R * π) / 180
L1 = (63°45'10" * 1200.00000 * π) / 180
L1 = **1335.23506**
L2 = (66°33'12" * 900.00000 * π) / 180

$L_2 = \mathbf{1045.41732}$

L.T.B. = Tangent In - Δ_1

L.T.B. = N55°23'11"W - 63°45'10"

L.T.B. at PRC = **N08°21'59"E**

Tangent Out = Tangent In - (Δ_1 - Δ_2)

Tangent Out = N55°23'11"W - (63°45'10" - 66°33'12")

Tangent Out = **N58°11'13"W**

$C = 2 * R * \sin(\Delta / 2)$

$C_1 = 2 * 1200.00000 * \sin(63°45'10" / 2)$

$C_1 = \mathbf{1267.41225}$

Chord Bearing 1 = Tangent In - (Δ_1 / 2)

Chord Bearing 1 = N55°23'11"W - (63°45'10" / 2)

Chord Bearing 1 = **N23°30'36"W**

$C_2 = 2 * 900.00000 * \sin(66°33'12" / 2)$

$C_2 = \mathbf{987.62831}$

Chord Bearing 2 = L.T.B. - (Δ_2 / 2)

Chord Bearing 2 = N08°21'59"E - (66°33'12" / 2)

Chord Bearing 2 = **N24°54'37"W**

Solution for Example 5 - Compound Curve:

Refer to Figure 5 for the following example:

Given:

$R_1 = 1000.00000$

$R_2 = 700.00000$

Δ1 = 63°45'10" RT
Δ2 = 40°34'32" RT
Tangent In = N53°25'41"W

Solve for the following elements:
T = R * Tan(Δ / 2)
T1 = 1000.00000 * Tan(63°45'10" / 2)
T1 = **621.87357**
T2 = 700.00000 * Tan(40°34'32" * 2)
T2 = **258.76812**
L = (Δ * R * π) / 180
L1 = (63°45'10" * 1000.00000 * π) / 180
L1 = **1112.69588**
L2 = (40°34'32" 700 * π) / 180
L2 = **495.72393**
L.T.B. = Tangent In - Δ1
L.T.B. = N53°25'41"W - 63°45'10"
L.T.B. at PCC = **N10°19'29"E**
Tangent Out = Tangent In - (Δ1 + Δ2)
Tangent Out = N53°25'41"W - (63°45'10" + 40°34'32")
Tangent Out = **N50°54'01"E**
C = 2 * R * Sin(Δ / 2)
C1 = 2 * 1000.00000 * Sin(63°45'10" / 2)
C1 = **1056.17687**
Chord Bearing 1 = Tangent In - (Δ1 / 2)

Chord Bearing 1 = N53°25'41"W - (63°45'10" / 2)

Chord Bearing 1 = **N21°33'06"W**

C2 = 2 * 700.00000 * Sin(40°34'32" / 2)

C2 = **485.42981**

Chord Bearing 2 = L.T.B. + (Δ2 / 2)

Chord Bearing 2 = N10°19'29"E + (40°34'32" / 2)

Chord Bearing 2 = **N30°36'45"E**

Solution for Example 6 - Boundary line along a Curve:

Refer to Figure 6 for the following example.

Given:

R = 2864.78898

L1 = 85.34'

L2 = 150.11'

Tangent In = N10°20'45"E

Solve for the following elements:

First solve for Δ1 & L.T.B.1:

Δ = (180 * L) / (R * π)

Δ1 = (180 * 85.34) / (2864.78898 * π)

Δ1 = **1°42'24" RT**

L.T.B.1 = Tangent In + Δ1

L.T.B.1 = N10°20'45"E + 1°42'24"

L.T.B.1 = **N12°03'09"E**

Next solve for Δ2 & L.T.B.2:

Δ2 = (180 * 150.11) / (2864.78898 * π)

Δ2 = **3°00'08" RT**

L.T.B.2 = L.T.B. 1 + Δ2

L.T.B.2 = N12°03'09"E + 3°00'08"

L.T.B.2 = **N15°03'17"E**

Next solve for C, Chord Bearing & Radial Bearings:

C = 2 * R * Sin((90 * L) / (π * R))

C = 2 * 2864.78898 * Sin((90 * 150.11) / (π * 2864.78898))

C = **150.09283**

Chord Bearing = L.T.B.1 + (Δ2 / 2)

Chord Bearing = N12°03'09"E + (3°00'08" / 2)

Chord Bearing = **N13°33'13"E**

Radial Bearing 1 = L.T.B.1 + 90°

Radial Bearing 1 = N12°03'09"E + 90°

Radial Bearing 1 = 102.05250°

Radial Bearing 1 = 180° - 102.05250°

Radial Bearing 1 = **S77°56'51"E**

Radial Bearing 2 = L.T.B.2 + 90°

Radial Bearing 2 = N15°03'17"E + 90°

Radial Bearing 2 = 105.05472°

Radial Bearing 2 = 180° - 105.05472°

Radial Bearing 2 = **S74°56'43"E**

Check:

Δ2 = Radial Bearing 1 - Radial Bearing 2

Δ2 = S77°56'51"E - S74°56'43"E

Δ2 = 3°00'00" (Equals previous calculation)

    Circular curves are not that difficult to solve once you master the basics. Drawing a sketch of the curve is very helpful in identifying the process needed to solve the problem at hand. This book has given you all of the tools that you will need to solve most circular curve problems that you will encounter in your career. Keep this book on your Kindle device or app for easy quick reference.

NOTES

ABOUT THE AUTHOR
Jim Crume P.L.S., M.S., CFedS

My land surveying career began several decades ago while attending Albuquerque Technical Vocational Institute in New Mexico and has traversed many states such as Alaska, Arizona, Utah and Wyoming. I am a Professional Land Surveyor in Arizona, Utah and Wyoming. I am an appointed United States Mineral Surveyor and a Bureau of Land Management (BLM) Certified Federal Surveyor. I have many years of computer programming experience related to surveying.

This book is dedicated to the many individuals that have helped shape my career. Especially my wife Cindy. She has been my biggest supporter. She has been my instrument person, accountant, advisor and my best friend. Without her, I would not be the professional I am today. Cindy, thank you very much.

Other titles by this author:

http://www.cc4w.net/ebooks.html

www.ingramcontent.com/pod-product-compliance
Lightning Source LLC
Chambersburg PA
CBHW041147180526
45159CB00002BB/743